YOUR KNOWLEDGE HAS VALUE

- We will publish your bachelor's and master's thesis, essays and papers

- Your own eBook and book - sold worldwide in all relevant shops

- Earn money with each sale

Upload your text at www.GRIN.com and publish for free

GRIN ☺

Pravin Gautam

Mountain Farming and Policies: Nepalese Perspective

GRIN Publishing

Bibliographic information published by the German National Library:

The German National Library lists this publication in the National Bibliography; detailed bibliographic data are available on the Internet at http://dnb.dnb.de .

Imprint:

Copyright © 2002 GRIN Verlag, Open Publishing GmbH
Print and binding: Books on Demand GmbH, Norderstedt Germany
ISBN: 978-3-640-99202-7

This book at GRIN:

http://www.grin.com/en/e-book/177250/mountain-farming-and-policies-nepalese-perspective

GRIN - Your knowledge has value

Since its foundation in 1998, GRIN has specialized in publishing academic texts by students, college teachers and other academics as e-book and printed book. The website www.grin.com is an ideal platform for presenting term papers, final papers, scientific essays, dissertations and specialist books.

Visit us on the internet:

http://www.grin.com/

http://www.facebook.com/grincom

http://www.twitter.com/grin_com

Mountain Farming and Policies:
Nepalese Perspective

Pravin Gautam

Geographical Condition

Nepal, predominantly a mountainous, agrarian and landlocked country, has an area of 147,181 sq.km. with a population of about 22.3 million. It is geographically divided into three geo-ecological zones: Mountain (4,877m – 8,848 m above the mean sea level), Hills (610 m – 4,877m), and Terai (59 m – 610m).

The Terai region, a low flat land, is an extension of the *Gangetic* plains of India. It is a narrow tropical belt occupying 23 percent of the country's total land area providing dwelling to 49.06 percent of the total population. About 40 percent area of the region is cultivable.

The Hill region consists of several peaks, fertile valleys and basins. 'Mahabharat Range', the largest hill range passes through the region. The region accounts for 42 percent of the country's land area. One tenth of the land area is suitable for cultivation. Around 44.3 percent of the country's total population resides in the region (NYB, 2001). Since the region comprised of broad complex of hills and valleys, very steep slope, even up to 30 degrees, is terraced for farming (Rajbhandari, 1999).

The mountain region covers about 35 percent of the total area of the kingdom, out of which only about two percent of the land is cultivable. Almost all major rivers of the country originate here. The region is consecrated with nine out of the 14 world's highest peaks that rise above 8000 meters. The region is thinly populated with 7.3 percent of the country's total population (CBS, 2001). The major occupation in the region is yak and sheep farming. Different dairy products are made from the milk of *chauri*. The high mountains are mainly used for grazing sheep and yak.

Besides the geo-ecological zones, administratively, Nepal is divided into five development regions, 14 zones and 75 districts. Sixteen districts are in the mountain region, 39 in the hill region and the remaining 20 districts cover the Terai region.

Socio-economic Life in the Nepalese Mountains

The importance of mountains for human subsistence cannot be overlooked. Mountains are regarded as a center for human civilization and cultural diversity. Mountains are the direct life-support base for people living in the region and also provide goods and services to the majority of the population. However, despite the value of mountains, mountain communities, especially mountain farmers, are experiencing acute socio-economic deprivation. The situation in Nepal is a suitable example for this.

The country occupies only about 0.1 percent of the total land area of the world, but it harbors a high proportion of the world's biodiversity; 2.6 percent of Algae, 3.45 percent of Pteridophytes, three percent of non-flowering plants, 2.6 percent of flowering plants, 2.65 percents of butterflies and moths, 9.3 percents of birds, and 4.5 percents of mammals (Regmi *et al.,* 1999). Further study on other groups is still lacking. Regarding agro-biodiversity, Nepal is a cradle of domestification and home to a diverse range of the world's cultivated species. In Nepal, over 500 plant species are found. Among which, 200 species are cultivated and 20 species are used as food (Gautam, 2002). Crops like rice, rice bean, eggplant, buckwheat, soybean, foxtail millet, citrus, mango etc. have high genetic diversity. Similarly, the diversity in the case of underutilized food crops and tropical fruit species is noteworthy. The mountain region accounts for almost one-third of the endemic species of the country, which in turn is about five percent of the total world flora. In terms of biodiversity, Nepal is the 25th richest country in the world and the 11th richest in Asia (Ghimire, 2002). These figures are impressive given Nepal's comparatively limited size.

Nevertheless, as yet Nepal has failed to realize this advantage and use it to benefit mountain communities. These natural wealth have not been able to simplify the poverty-stricken life of mountain farmers. There is a vast variation in the poverty across the three ecological zones (Table 1).

Table 1.   Incidence of Poverty by Geographical Region in Nepal

Geographical region	Percentage of households below poverty line			Percentage pf population below poverty line		
	Rural	*Urban*	*National*	*Rural*	*Urban*	*National*
Terai	34.1	20.2	33.0	35.4	24.1	34.5
Hills	49.8	12.6	47.1	52.7	14.5	50.0
Mountain	36.0	-	36.0	44.0	-	44.1
Source: Rajbhandari, 1999.						

Social indicators remain low for the vast majority of the people, and are especially low for the marginal mountain farmers. Access to social services such as health, education, and drinking water is much more limited for mountain communities compared to their counterparts living in the Terai. On the other hand, human poverty index is the highest in the mountains followed by the hills. Out of the total 75 districts, 60 are food insecure and most of these are in the hill and mountain regions. Further difficult terrain, rugged topography, lack of transportation facilities, limited cultivable land, and unfavorable climatic conditions are major problems in the mountains of Nepal. It is noteworthy to mention that 20 districts of the hill and mountain regions have yet to be linked by road (Ghimire 2002). The land distribution pattern is highly skewed. About 75.6 percent of the household owns agricultural land. The top six percent of the population have control over 33 percent of all agricultural land. Similarly about 69.5 percent of land holdings are less then one hectares. There is relatively smaller size of land holdings in the hills (0.77 ha) and mountains (0.68 ha) compared to the average size of landholding in the Terai (1.26 ha). In addition to this, agricultural lands in the mountains and hills are less fertile and less productive. They are fragile and vulnerable to land slide and erosion loosing fertile topsoil.

Similarly, the demands of agricultural labor are highly seasonal and, especially in the hills and mountains, there are very few opportunities for non-farm activities. It is estimated that, on average, a farm worker in Nepal is employed only for 55 days per year in hilly regions.

In addition, due to poverty and disparities, mountains are becoming the center for anti-State activities. At present, around 82% of these activities in the world are operated from mountains (Shrestha, Tirtha Bahadur in Ghimire, 2002). Nepal has been facing the same problem for the past six years. This has further deteriorated the socio-economic condition of the mountain farmers.

Nevertheless, Nepal's mountains and hills need not be problematic. There are also several opportunities. Herbal plants found in mountains and hills have direct positive impact on the economic life of mountain people. About 690 types of herbs are found in Nepal and almost 100 have them commercial value (Table 2). These herbal plants are found up to 8000 m. altitude.

Table 2.   Some Herbal Plants Found in Hills/Mountains of Nepal having Commercial Value

Soap pod (*Acacia rugata*), Sweet flag (*Acorus calamus*), Cinnamon (*Cinnamomum zeylanicum*), Cedar, Utrasum bead tree (*Elaeocarpus sphaericus*), Padam Chal (*Rheum webbianum*), Soap-nut tree (*Sapindus mukorossi*), Orchis (*Dactylorhiza hatagirea*), Yarsa gumba (*Cordyceps sinensis*), Stinging nettle (*Urtica diocia*), Neem (*Azadirachta indica*), Nicker bean (*Entada phaseoloides*), Barbery (*Berberis aristata*), Emblic myrobalan (*Emblica officinalis*), Ephendrine (*Ephendra gerardiana*), Madder (*Rubia manjith*), Gurjo (*Tinispora cordifilia*), Oriental cashewnut (*Semecarpus anacardium*), Chiretta (*Swertia chirayita*), Common dandelion (*Taraxacum officinale*), Himalayan silver fir (*Abies spectabilis*), Himalayan birch (*Betula utilis*), White sandle wood (*Santalum album*), Rockfoil (*Berginia chiliata*), Valerian (*Valeriana jatamansii*), Club moss (*Lycopodium clavatum*), Spikenard (*Nardostachys grandiflora*), Common Pursulane (*Portulea oleracea*), Long peeper (*Piper longum*), Bastard myrobalan (*Terminalia bellirica*), Kutkee (*Picrorhiza scrophulariiflora*), Nepal cardamom (*Amomum aromaticum*), Asparagus (*Asparagus racemosus*), Indine blue pine (*Pinus wallichiana*) etc.

Similarly, mountains and hills also provide tremendous opportunity for the cultivation of cash crops and high-value crops. These areas are suitable for the cultivation of crops like tea, cardamom, ginger and mushroom. Similarly, mountains have suitable climatic condition for the production of vegetable seeds as well as off-season vegetables. In addition to this, hills and mountains also have floriculture and tourism opportunities.

Agricultural Aspect
Agriculture is the mainstay of the Nepalese economy. As per revised estimates, it accounted for 37.88 percent in national GDP in the year 2000/2001. In Nepal, total agricultural land is 2392.90 thousand hectares. Out of total arable land (2,323.40 thousand hectares) 6.9 percent and 37.5 percent land is in hills and mountains. Similarly, out of total area of holdings, only 23.6 percent of the land area in mountains and 23.4 percent in hills have irrigation facilities (CBS, 2002).

The foothills of mountains are suitable for cultivating paddy, maize, barley, buckwheat etc. In addition to this potato, cauliflower, cabbage, beans and low temperate fruits are also cultivated in the region. Paddy, wheat, maize, millet, barley, buckwheat are the major cereal crops cultivated in hills. Different fruits and vegetables are also cultivated. There has been no significant mechanization in farming due to inaccessibility and the geographical condition in hills and mountain areas.

Table 3.   Area Under and Estimated Production of Major Food & Cash Crops in Hills cum Mountains of Nepal

Food	1997/98		1998/99		1999/00		2000/01	
Crops	A	P	A	P	A	P	A	P
Paddy	408220	897890	417284	974428	423656	964999	431931	1035093
Maize	624200	1031980	625140	1358162	646360	1115389	656526	1161347
Wheat	280290	665	283431	1184	288912	448507	289953	461429
Barley	35140	35370	29750	29658	26865	29483	26799	29142
Millet	247350	270020	248208	274839	252555	258639	249059	271224
Cash Crops								
Sugarcane	1770	38360	1910	1910	1694	38448	1737	39615
Oil Seed	37800	23970	37303	26820	36976	25560	36984	26254
Tobacco	210	140	93	33	48	29	31	25
Potato	77150	660150	78989	697855	80596	734008	82368	773170

Source: CBS, 2002   *A: Area in Hectare; P: Productioon in Metric ton*

Livestock is an integral part of the hill/mountain farming in Nepal. Livestock is the source of milk, meat, manure, draft power and household income. Poultry raising is the rising component of hill/mountain farming. As livestock are integral part of hill/mountain farming, the majority of whom are subsistence mountain farmers, it must be maintained at a certain minimum threshold. This has resulted in heavy pressure on maintaining livestock. The population of livestock in Nepal in 2000/01 was about 6.9 million head of cattle, 3.6 million of buffaloes, 6.4 million of goats, 0.85 million sheep, 0.91 million pigs, 19.7 million fowl and 0.41 million ducks (CBS, 2002). Livestock contribute about 20 percent of household cash income in the hills and mountains, without considering home consumption of livestock products into account (Tulachan and Neupane, ____)

Table 4.   Livestock Population and Composition in Hills and Mountains of Nepal

Livestock classes	Mountains		Hills	
	Change in population (1988/89-1996/97)	Change in share* (1988/89-1996/97)	Change in population (1988/89-1996/97)	Change in share (1988/89-1996/97)
Cattle	+ 3.17	+ 0.89	+ 5.77	- 0.51
Buffaloes	+ 0.58	0	+ 8.30	+ 0.21
Sheep	- 9.59	- 1.70	- 2.53	- 9.59
Goat	+ 2.87	+ 0.80	+ 9.37	+ 2.87
Source: Tulachan and Neupane,_____.		* Percentage share of individual species in total livestock population (Cattle+ Buffaloes+Sheep+Goats)		

Policy Regime

Although historically, planning in Nepal started in 1935 AD when Udyog Parishad (Industrial Board) was established and the 20-year plan was announced after the Second World War, actual planned development started only in 1956 when the first five-year plan was launched. Many previous attempts had failed because of political instability in the country. The agriculture sector was prioritized in all five-year plans. However, it was only in 1975 that the agriculture sector became the primary priority. Since then it has been accorded high priority in all five years plans with the aim of achieving broad-based growth.

The paramount objective of the planned development efforts in Nepal is poverty alleviation. The Ninth Plan (1997-2002) established a 20 years timeframe for poverty alleviation. The recently adopted approach paper for the Tenth Plan also expresses commitments to reduce the level of poverty in the country through agricultural development. For the past 12 years Nepal has been moving towards economic reform through liberalization. Being an aid-dependent country, its development policies should harmonize with donors. In view of that, a priority reform program has been adopted. Like other sectors of the economy, agriculture has been liberalized and the elements of liberalization includes:

- De-controlling prices of agricultural inputs.
- Withdrawal of irrigation subsidy on shallow tube-wells, transportation subsidy of chemical fertilizer and food grains.
- Privatization of veterinary services.
- De-monopolization of supply of seeds.
- Establishment of national agricultural development fund for competitive grants for research and development activities.
- Creation of marketing centers.
- Partnership and contract based extension services through private service provider.

The 20-year Agricultural Perspective Plan (APP) was implemented in 1995 which is an agriculture-led growth strategy for poverty reduction and overall economic development of the country. It aims to increase the agricultural growth from the level of 2.5 percent to five percent. It also emphasizes that commercially oriented agriculture has a significant impact both in terms of increased income and in terms of the environment. The plan has three focus areas: crop intensification, high growth of agriculture for employment and agriculture as a lead sector for economic transformation. Rural roads and power have been prioritized as inputs along with fertilizer, irrigation and technology. Similarly, livestock, high-value crops, agribusiness and forestry have been prioritized as outputs. The plan has identified different strategies for Terai and hills/mountains. The strategy for the hills and mountains is demand-driven in contrast to the technology-driven strategy in Terai. The plan has, however, no reference to land use policy, which is a significant factor in land management.

The latest in a series of long term plans from the Nepalese government is 'Vision 2020' which aims to boost the technical aspects of the agriculture sector. The policy particularly concentrates on technology generation, development and their efficient and effective use in agriculture and related natural resources. Market analysis, irrigation management, agro-forestry and policy research are among the core components of the policy.

Evaluation of Past Policies

Despite numerous plans and policy efforts, the performance of the agriculture sector is still dwindling. While Nepal's development efforts have been hindered by its largely rugged topography, highly skewed distribution of limited natural resources, rapidly increasing population and limited development infrastructure, there are many other obstacle that have hindered the effective implementation of the plans. This has resulted in slow growth in the production of food grains leading to the weakening of the food security situation and the economic base of farming communities especially in mountains. Other realities and negative impacts of past plans and policies can be summarized as follows:

- Land fragmentation has increased making the land unsuitable for mechanization.
- Monoculture has replaced the mixed and other traditional cropping pattern leading to the loss of genetic diversity.
- Very steep slopes and marginal lands are being shifted to agriculture in order to meet the needs of a population increasing at alarming rate. This has resulted in low productivity.
- Natural forest and other natural resources were over utilized and this has resulted in the increased landslides, erosions, drought and flooding every year.
- Food security situation is becoming more acute on a daily basis and 60 districts are food insecure and majority of which are in hills and mountains. Hundreds of people die due to hunger in far western hills and mountain every year.
- Physio-chemical properties of agricultural land have been degraded due to imbalance and excessive use of chemical fertilizer.
- Though the production of major food crops has increased, it is mainly as a result of increased cropping area rather than increased technology and productivity.
- About 66 percent of land has yet to have an irrigation facility.
- Nepal has shifted from net food exporter to net food importer since the 1980's.
- Sluggish agricultural development, land shortage and population growth have pushed more and more families into landlessness.

Mountain Development Issues in Policies

The suffering of the mountain communities is gradually increasing and their standard of living is declining because they have been neglected at both policy and practice levels by the governments. Despite the fact that the Nepal is a predominantly mountainous country, it has no mountain specific development policies. Even the Ninth plan did not address mountain development specifically.

However, it is not that there have not been policy initiatives for mountain development. In the past, some policies were formulated to cover mountain development. The concept of Regional Planning during 1970's can be regarded as an example, which covered the hills and mountains districts. But the problem was in faithful implementation. Government's development policies, which somehow related to the mountains, are of four types. Firstly, regular government programs implemented by line agencies at the local level, which basically focus on local and community development with social services. Secondly, local development programs to be implemented by local government institutions. Thirdly, plans and policy targeted to develop rural and remote areas, which covers the remote areas of mountain and certain communities. And lastly, the Integrated Rural Development Program, which was directed at improving social services and establishing integrated ways for infrastructure development. However, these plans and programs have not been able to improve the situation of mountain farming communities. As aforesaid, these plans and policy are often adapted to comply with donor policy. Thus, these programs usually don't have sustainability. APP can be taken as a suitable example for this. APP has identified the comparative advantage strategy for mountain development. Nonetheless, this strategy of comparative advantage has not received attention in resource management and distribution.

Other plans and policy for mountain resource management are: Forest Act 1993, Water Resource Act 1992, National Resource Conservation Strategy 1989, Environment Protection Act 1997, Environmental Policy and Action Plan 1993, and Natural Parks and Wildlife Conservation Act 1973. These have prioritized the conservation of natural resources in general and specifically the conservation of forest, wildlife and water bodies. However, the needs of mountain peoples have been overlooked. This has created social problems and conflict. Government has therefore responded by changing its approach and has tried to create greater economic opportunities for mountain communities.

Farmers' Rights, World Trade Organization (WTO) and Nepal
Farmers' rights, as per the Food and Agriculture Organization (FAO), acknowledge the contribution of farmers in conservation and development of plant and animal genetic resources. To the point, this concept aims to protect the farmers against International Union for the Protection of New Varieties of Plants (UPOV), an international agreement signed in 1961. Nepal, yet, has not joined UPOV and also don't have legal apparatus to protect plants breeders rights on new varieties. Similarly, there are not farmers' rights specific laws in Nepal. There are, nonetheless, some legislative instruments related to farmers' rights. Important acts and policies somehow related to farmers' rights are: Agriculture Perspective Plan (1997-2017), Environment Protection Act 1997, Seed Act 1988, Forest Act 1995, Plant Protection Act 1972, National Parks and Wildlife Conservation Act 1973, Land Act 1964, Plant Quarantine Act 1972, Nepal Agricultural Research Council Act 1992, Ninth Plan (1997-2002), Water Resource Act 1992, Aquatic Animal Protection Act 1961, Livestock Health and Services Act 1998 etc.

Nepal had applied for the membership to General Agreement on Trade and Tariff (GATT) in 1989 and got the status of Observer after the transformation of GATT to World Trade Organization (WTO). Accession negotiations are going on and government has said that Nepal will obtain WTO membership within one to one and half year. However, particularly under the WTO agreements like Trade Related Intellectual Property Rights (TRIPs), Agreement on Agriculture (AoA), Sanitary and Phytosanitary (SPS) Measures etc, farming communities in mountain areas are in risk of being deprived of their traditional knowledge system. As the country has no rigid legislative measures to check bio-theft, farmers' right to produce, store, reuse and exchange their plant and seed varieties is being abducted. Monopolies and patents on life forms by the multinational corporations have jeopardized their livelihoods.

Against the centuries of tradition of seed ownership and the liberty to use them, global patenting rules now threaten the livelihoods of these vulnerable mountain-farming communities. The TRIPs Agreement violates the farmers' rights by allowing countries to provide patent protection for microbiological and non-biological processes, microbes, and plant varieties. Nepal being at the low end of the spectrum of generation and use of intellectual property, it has not yet enacted laws covering these aspects. Until now, a total of 44 patent and 29 design protections have been granted and out of them 17 patents and 24 design protections were of foreign origin. In fact, Nepal's IPR regime is different from TRIPs. In the Nepalese context, geographical indication and trade secrets are excluded from protection under the current laws. Similarly, in terms of flexibility there is not a provision for compulsory licensing, parallel imports and *sui generis* protection. Patenting is of seven years and poor institutional responses and negligible punishments compromise enforcement.

The AoA under WTO directly affects farming communities in mountainous areas because subsidies are to be reduced or abolished under this agreement and the Nepal has been practicing the same as a way to liberalized economy. Although these subsidies in agriculture have distorted the pattern of resource use, they have been important for the initial development of inaccessible remote areas in hills and mountains. Due to geological condition and steep terrain in western hills and mountains, food has to be transported through sky route from city centers during the period of food scarcity every year. In absence of the government's ability to subsidize in transportation, hundreds of poor and marginal people die due to hunger. Across developing countries including Nepal, subsidies are being reduced or abolished whilst overall subsidization of agricultural producers in the OECD countries is much higher today than it was in 1995. The other problem with trade liberalization policies is that they place a great emphasis on cash crops for export. Consequently, agricultural lands are increasingly being devoted to the production of mass scale monoculture cash crops at the expense of food crops. This aggravates the already food-insecure situation.

The traditional or indigenous food plants have different advantages (like specifically designed to work with the monsoon weather pattern and maturity needs; adoption to local soil, climate and growing conditions; low risk of total crop failure; natural resistance/tolerance to different biotic and abiotic stresses prevailing in the growing area etc) over newer varieties. All these are essential components of sustainable agriculture but they are fast vanishing. The condition will further aggravate under different agreements of WTO.

Along with the challenges, there are also opportunities which globalization presents. However, despite the opportunities like enhanced market access for Nepal's export products in the international market, export diversification that includes products from mountains, the actual gain to Nepal will depend on its capacity to formulate and pursue development strategies for mountain agriculture under the WTO framework.

Conclusion
It is the fact that hills/mountains agriculture is pivotal in the Nepalese economy. However, hills and mountain communities are deprived of most of the basic human needs. These areas are fragile and vulnerable to landslides and erosion leading to the degradation of these hills and mountains and depletion of natural resource, biodiversity. Poverty, unemployment and other disparities are chronic in these areas. The problem is further impounded by Nepal's locational disadvantage which make it almost impossible to make basic infrastructure facilities available to hill and mountain communities. In addition, the mountain areas are ignored during the policy making

process. Nonetheless, hills and mountains cannot be seen as necessarily problematic. In fact, there are several opportunities like, richness in biodiversity, herbal and medicinal plants farming, tourism, different high-valued crops etc., in hills and mountain areas which can be used for their overall development if proper plans and policies are formulated and implemented faithfully in order to utilize these natural resources.

It seems that Nepal has no alternative than to concentrate its economic resources in the agricultural sector to receive the necessary benefits in the context of liberalization and globalization. Whether Nepal joins the WTO or not, different agreements of this global body will impact either directly or indirect on its agriculture. If domestic plans and policies are not amended in light of the changing economic environment, Nepal will certainly pay in terms of biodiversity loss, dumping of cheap products, devastation of domestic agriculture and agro-based industry due to lack of ability to compete with big corporations. Therefore, existing plans and policies should be amended and new laws and acts should be enacted, if necessary, to protect Nepalese agriculture in general and mountain agriculture in particular.

Suggestions

- Mountain specific development plans and policies should be formulated taking into account the socio-economic, cultural and geological conditions in the hills and mountains.
- High-value crops and crops for which Nepal has a comparative advantage like cardamom, orthodox tea, zinger, milk and milk products, spices, vegetables seed etc. should be promoted for the upliftment of the economic condition of mountain farmers.
- Agricultural planning and backup packages should be based on existing farming system in hills and mountains rather than introducing new ones.
- Past trends of growing everything everywhere should be changed and policy should recognize several agro-ecological sub-regions with different potential for specific agricultural developments.
- Forest, livestock and agriculture programs should be implemented in integrated ways.
- Better access to improved inputs for increased food production; together with training for wage employment opportunities should be provided to marginalized mountain farmers.
- Appropriate policy should be formulated in order to utilize the natural resources efficiently for mountain development.
- Sufficient road networks should be developed.
- Strong linkage should be developed between mountain/hills and Terai.
- Plans and policies should be formulated that aim to conserve and promote the indigenous plant and animal varieties along with the knowledge base of farming community to utilize, conserve and promote them.
- Production base should be diversified raising the output level to exploit the opportunities given by the secured and predictable market environment.
- Increased access to farmers, in particular those farmers from indigenous groups of the mountain and hill regions, to agricultural training, credit and use of improved technology should be provided to ensure food security.
- Farmers' Rights should be clearly defined and separate acts should be formulated granting the farmers the rights to save seeds and exchange or sell them.
- Appropriate legal mechanism for the protection of biodiversity should be constructed and biodiversity registration should be started at community level as soon as possible to safeguard these communities from the negative impacts of the TRIPs agreement.

- Any future investment in the mountains must ensure sustainable human development so that the food security and livelihood of the farmers and local producers can be ensured.

References

Anonymus. _____. Glossary of Some Important Plants and Animal Names in Nepal.

Bhattarai, Chiranjibi. 1999. Role of Local Institutions in Biodiversity Conservation. Pro Public, Kathmandu, Nepal.

CBS. 2002. Statistical Pocket Book Nepal 2002. National Planning Commission Secretariat, Cental Bureau of statistics, Kathmandu, Nepal.

CBS. 2002. Population Census 2001: National report. National Planning Commission Secretariat, Cental Bureau of statistics, Kathmandu, Nepal.

Chaudhary, Moushumi. 2002. Draft Policy Documents for Asia. Asia High Summit, 6 –10 May 2002, Kathmandu, Nepal.

Dahal, Madan K (Ed). 2001. Future of Nepalese Economy. Nepal Foundation for Advanced Studies (NEFAS), Kathmandu and Friedrich-Ebert-Stiftung (FES), Kathmandu, Nepal.

Gautam, Pravin. 2002. Biodiversity Conservation: Need to be serious. The Rising Nepal, 03 June 2002, Kathmandu, Nepal.

Gautam, Pravin. 2002. World Food Summit: Will it Come Up with Solutions? The Space Times Today, 12 June 2002, Kathmandu, Nepal.

Ghimire, Hiramani. 2002. *"Pahadi Chetra Ka Bishesh Awasaryukta Utpadanharu: Krishak ko Aajiwika"*. Pro Public, Kathmandu, Nepal.

Ghimire, Hiramani. 2002. Farmers' Rights and Mountain Communities in Nepal. SAWTEE, Kathmandu, Nepal.

Gurung, Harka. 2002. Policies for Mountain Development: Evidence from the Hindu Kush – Himalaya – Hengduan. Asia High Summit, 6 –10 May 2002, Kathmandu, Nepal.

Jodha, N.S. 2002. Policies for Sustainable Mountain Development: An Indicative Framework and Evidence. Asia High Summit, 6 –10 May 2002, Kathmandu, Nepal.

Munir, Shafqat. 2002. Mountain Issues and Farmers' Rights: South Asian Perspective. Trade and development Monitor, BGMS 2002 Special Issue, SAWTEE, Kathmandu, Nepal.

Nathan, Dev. 2002. Comparative Advantage and the Development of Mountain Communities in a Globalising Capitalist System. Asia High Summit, 6 –10 May 2002, Kathmandu, Nepal.

Nepal Year Book. 2001. Ed: Ramesh C Arya, Arunodaya Traders Coop. Ltd., Kathmandu, Nepal.

Pandey, Posh Raj. 1999. World Trade Organisation and Nepal: Opportunities and Challenges in *WTO: Regional Cooperation and Nepal* (Ed: Horst Mund). Nepal Foundation for Advanced Studies (NEFAS), Kathmandu and Friedrich-Ebert-Stiftung (FES), Kathmandu, Nepal.

Rajbhandari, Binayak. 1999. Food Security and Sustainable Livelihoods: The Bio-Intensive Farming Innovation against Hunger in the Nepalese Context. WOREC, Kathmandu, Nepal.

SAWTEE. 2002. IYM 2002 Special Issue. Kathmandu, Nepal.

Tulachan, Pradeep Man and Arun Neupane. _____. Livestock in Mixed Farming Systems of the Hindu Kush-Homalayas: Trends and Sustainability. Food and Agriculture Organisation (FAO) and International Center for Integrated Mountain Development (ICIMOD).